W9-CDJ-785

EVERYDAY SCIENCE

Changing Forms

Please visit our web site at: www.garethstevens.com
For a free color catalog describing Gareth Stevens Publishing's list of high-quality books
and multimedia programs, call 1-800-542-2595 or fax your request to (414) 332-3567.

Library of Congress Cataloging-In-Publication Data

Riley, Peter D.
    Changing forms / by Peter Riley. — North American ed.
        p. cm. — (Everyday science)
    Summary: Describes different kinds of materials and uses simple experiments to
explain how their shape or state of being can be changed.
    Includes bibliographical references and index.
    ISBN 0-8368-3246-9 (lib. bdg.)
    1. Change of state (Physics)—Juvenile literature. 2. Elasticity—Juvenile literature.
[1. Change of state (Physics)—Experiments. 2. Experiments.] I. Title.
QC301.R54   2002
530.4'74—dc21
                                                              2002022632

This North American edition first published in 2002 by
Gareth Stevens Publishing
A World Almanac Education Group Company
330 West Olive Street, Suite 100
Milwaukee, Wisconsin 53212 USA

Original text © 2001 by Peter Riley. Images © 2001 by Franklin Watts.
First published in 2001 by Franklin Watts, 96 Leonard Street, London, EC2A 4XD, England.
This U.S. edition © 2002 by Gareth Stevens, Inc.

Series Editor: Rachel Cooke
Designers: Jason Anscomb, Michael Leamen Design Partnership
Photography: Ray Moller (unless otherwise credited)
Gareth Stevens Editor: Lizz Baldwin

Picture Credits: Pictor International, p. 22 (b).

The original publisher thanks the following children for modeling for this book: Jordan Conn, Nicola Freeman, Charley Gibbens,
Alex Jordan, Eddie Lengthorn, and Rachael Moodley.

Printed in Hong Kong

1 2 3 4 5 6 7 8 9 06 05 04 03 02

EVERYDAY SCIENCE

Changing Forms

Written by Peter Riley

Gareth Stevens Publishing
A WORLD ALMANAC EDUCATION GROUP COMPANY

About This Book

Everyday Science is designed to encourage children to think about their everyday world in a scientific way, by examining cause and effect through close observation and discussing what they have seen. Here are some tips to help you get the most from **Changing Forms**.

• This book introduces the basic concepts of how materials change form and some of the vocabulary associated with them, such as melting, cooking, and the comparison of solid and liquid, and it prepares children for more advanced learning about materials changing forms.

• On pages 15, 17, 19, and 25, children are asked to predict the results of a particular action or activity. Be sure to discuss the reasons for any answers they give before turning the page. Most of these activities have only one possible result. Discuss the reasons for each result, then create other activities for the children and discuss possible outcomes.

• Encourage children to think about how materials change form by drawing on everyday experiences, particularly with food. Using the melting butter test on pages 17 and 18, for example, ask children questions such as: How could they make the butter melt faster? Will the butter always melt? What else happens to the butter? Does the butter change color?

• Relate the discussions on pages 8 through 12, about changing the shapes of materials, to using forces such as pushing and pulling.

• Link the way materials change, or do not change, to the choice of materials for particular purposes. On page 10, for example, would a headband made of material that does not stretch be as useful? On page 19, would a metal fork be useful if it melted in sunlight?

Contents

Different Materials   6

Changing Shape   8

Stretching   10

Twisting   11

Bend It! Squash It!   12

Baking Hot   14

Cooking Food   16

Melting Materials   18

Freezing Liquids   20

Ice and Steam   22

Dissolving Materials   24

Test Results   26

Useful Words   28

Some Answers   29

For More Information   30

Index   31

Different Materials

Everything is made of materials.

wool

wood

wax

Materials can change in many different ways.

sand

dough

Some materials can change shape.

clay

Clay can be changed into lots of shapes.

6

Some materials change when they get wet.

When dry sand gets wet, the grains of sand stick together.

Some materials change when they get hot.

When a candle burns, the heat of the flame melts the wax.

Can you think of other ways materials change?
What are they?

Changing Shape

You can change the shape of some materials by stretching them.

When you blow up a balloon, you stretch it.

You can change the shape of some materials by twisting them.

When you tie string into a bow, you twist it.

Look closely at a piece of string. How is it made?

You can change the shape of some materials by bending them.

You can bend a plastic straw.

You can change the shape of some materials by squashing them.

You can squash an empty drink carton.

Try changing the shape of a sponge by stretching it, twisting it, bending it, and squashing it.

What happens when you let go of the sponge?

Stretching

When you pull on both ends of a lump of clay, you stretch the clay.

Lucy's headband stretches to keep her hair in place.

Twisting

When you turn the ends of two rolls of clay in opposite directions, you twist the clay.

This pasta has been twisted into spirals.

How is this roll of clay being changed? Turn the page to find out.

Squash It!

The clay is being bent.

Lots of materials can be changed by bending them.

Paper can be bent, or folded, to make an airplane.

Materials can also be changed by pushing down on them to squash them.

Lucy is squashing dough with a rolling pin.

Collect some different materials. Try to stretch, twist, bend, and squash them to find out how many ways they can be changed.

Make a table to show your results.

	stretch	twist	bend	squash
rubberband	✓	✓	✓	✓
plastic ruler			✓	
paper bag		✓	✓	✓
aluminum can			✓	✓
wool sock	✓	✓	✓	✓

Baking Hot

Toby is making a figure out of clay.

He changes the shape of the clay by stretching it and by squashing it.

When an adult puts the clay figure into a very hot oven, the heat of the oven changes the clay, too.

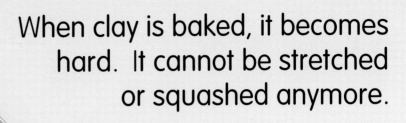

When clay is baked, it becomes hard. It cannot be stretched or squashed anymore.

But baked clay can be changed in a different way.

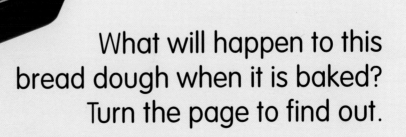

If Toby drops his clay figure, it will break into pieces.

What will happen to this bread dough when it is baked? Turn the page to find out.

Cooking Food

Baking changes the dough to bread.
It becomes hard on the outside
but stays soft inside.

Food changes when it is
heated. Heat cooks food.

Bacon is cooked
in a frying pan.

Slices of bread are
heated in a toaster.

Heat changes bacon. Before cooking, the bacon is soft and greasy. After cooking, it is hard and crispy.

Heat changes bread, too. The heat of a toaster makes bread hard and brown.

Lucy is spreading butter on some toasted bread. What happens to the butter? Turn the page to find out.

Melting Materials

The butter melts!

Try this test to learn more about melting. Gather the objects pictured below and place them on a tray or a large plate.

plastic cup

wax candle

metal fork

Leave them in a sunny place for three hours to see how heat from the Sun changes them.

chocolate bar

ice cubes

The metal fork and the plastic cup become warm.

The wax candle becomes soft.

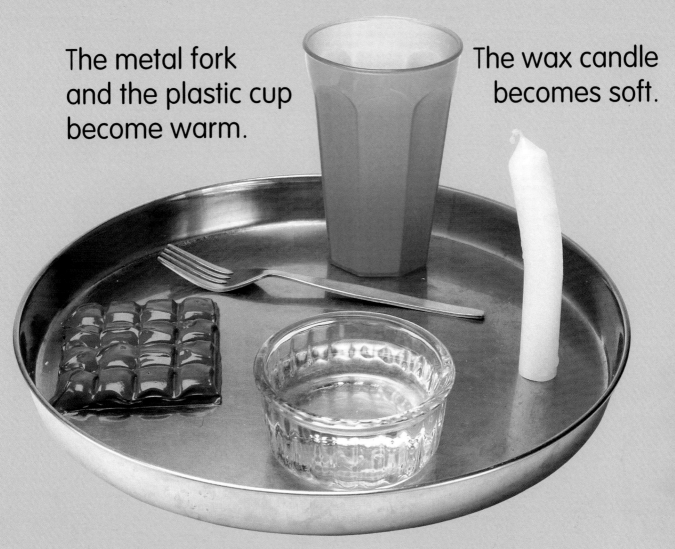

The chocolate bar melts.

The ice melts to water.

What will happen if you put the water into a freezer? Turn the page to find out.

Freezing  Liquids

The water freezes to solid ice!

Water, tomato sauce, and orange juice are liquids. When you pour them they flow.

Pour some water, tomato sauce, and orange juice into an ice cube tray. Put the tray in a freezer for about three hours.

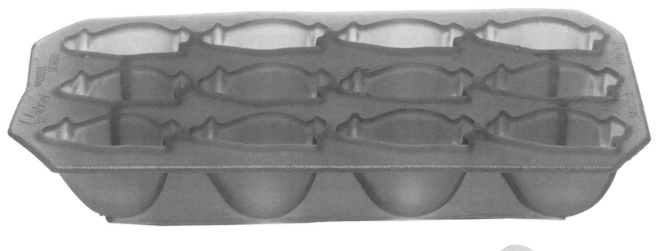

When you take the tray out of the freezer, the water, tomato sauce, and orange juice are hard. None of them flow anymore. They have changed from liquids to solids.

How can you change these solids back to liquids?

Both cold and heat can change the form of water.

Cold can freeze water, turning it to ice.

In very cold weather, dripping water freezes into icicles.

Heat can make water boiling hot.
You can see bubbles in boiling water.

Boiling water
changes into very
hot steam. Steam
is a kind of gas.

Steam made
this mirror misty.

If you draw a picture
in the mist with your
finger, your finger will
get wet. The steam
has changed back
into water.

Dissolving Materials

When you mix sugar with a warm liquid, the sugar disappears. This change is called dissolving.

Some materials, such as instant coffee, color the water as they dissolve.

If a material does not disappear in water, it does not dissolve.

Nadia is pouring warm water into six jars.

She stirs instant coffee, bath salts, flour, oil, sugar, and sand into different jars.

What do you think happens? Turn the page to find out.

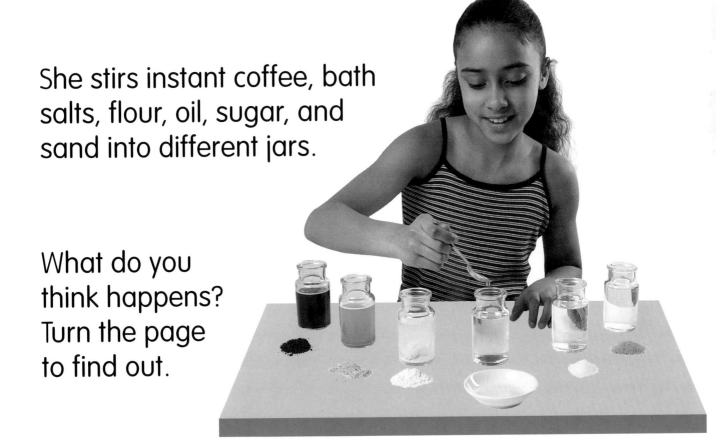

The instant coffee,
bath salts, and
sugar dissolve.

The oil, flour, and sand do not dissolve.

Nadia records her results in a table.

	dissolved	did not dissolve	other changes
instant coffee	✓		turned water brown
bath salts	✓		turned water blue
flour		✓	became very lumpy
oil		✓	floated on surface
sugar	✓		no other change
sand		✓	grains sank to bottom of jar

Do any other materials dissolve completely in water like the sugar did? How can you tell that there is sugar in the water?

Useful Words

cook: to prepare food for eating by heating it in some way.

dissolve: to mix a solid and a liquid so that the solid seems to disappear and become part of the liquid.

freeze: to change from a liquid to a solid by cooling the liquid at a very low temperature.

gas: a state of matter that is neither a solid nor a liquid. A gas does not have a shape, and it is often invisible.

ice: water that is frozen solid.

liquid: a material, such as oil or water, that has no shape of its own, and it flows.

materials: types of solid matter, such as wood, metal, or plastic, that are used to make objects.

melt: to change from a solid to a liquid by using heat to warm the solid.

solid: a material that has its own shape and does not flow.

steam: a gas that forms when water is heated until it boils.

Some Answers

Here are some answers to the questions asked in this book. If you had different answers, you may be right, too. Talk over your answers with other people and see if you can explain why they are right.

page 7   This book explains a number of different ways materials change. Some materials change when they are cooled. Sometimes heat will melt a material, but sometimes it will burn it. Some materials change when they are mixed with other materials.

page 8   String is made by twisting several thin strands of thread together. Look at string made of different materials, such as wool, cotton, and silk, to see how it is made.

page 9   When you let go of a sponge after stretching, twisting, bending, or squashing it, the sponge springs back to its original size and shape.

page 21   You can change these solids back to liquids by heating them. They will also melt if you just leave them out of the freezer for a while. You could speed up the melting by putting them in a sunny place. Don't forget to put the solids on a tray or a plate! Watch to see if they all melt in the same amount of time.

page 27   Like sugar, table salt also dissolves completely in water. The coffee and the bath salts dissolved completely, but they did not disappear in the water. They turned the water a different color when they were added. You can tell that there is sugar in it by tasting the water.

For More Information

More Books to Read

- *The Science of Liquids & Solids. Living Science* (series)
  Krista McLuskey
  (Gareth Stevens Publishing)

- *Solids, Liquids, and Gases. Starting with Science* (series)
  The Ontario Science Centre
  (Kids Can Press)

- *Solids, Liquids, Gases. Simply Science* (series)
  Charnan Simon
  (Compass Point Books)

Web Sites

- BrainPOP: States of Matter
  www.brainpop.com/science/matter/statesofmatter

- Science Rocks! Keep-a-Cube
  pbskids.org/zoom/sci/keepcube.html

Index

baking 14-15, 16
bending 9, 12, 13
boiling 23

changing shape 6, 8-9, 14
cooking 16-17

dissolving 24, 26, 27

freezing 19, 20-21, 22

heating 7, 14, 16, 17, 18, 22, 23

ice 18, 19, 20, 22

liquids 20, 21, 24

materials 6-7, 8, 9, 12, 13, 24, 27
melting 7, 18-19

solids 20, 21
squashing 9, 12, 13, 14, 15
steam 22, 23
stretching 8, 9, 10, 13, 14, 15

twisting 8, 9, 11, 13

water 19, 20, 21, 22, 23, 24, 25, 27